P9-CQJ-485

0 00 30 0429049 4

Energy in Action™

SOLAR ENERGY

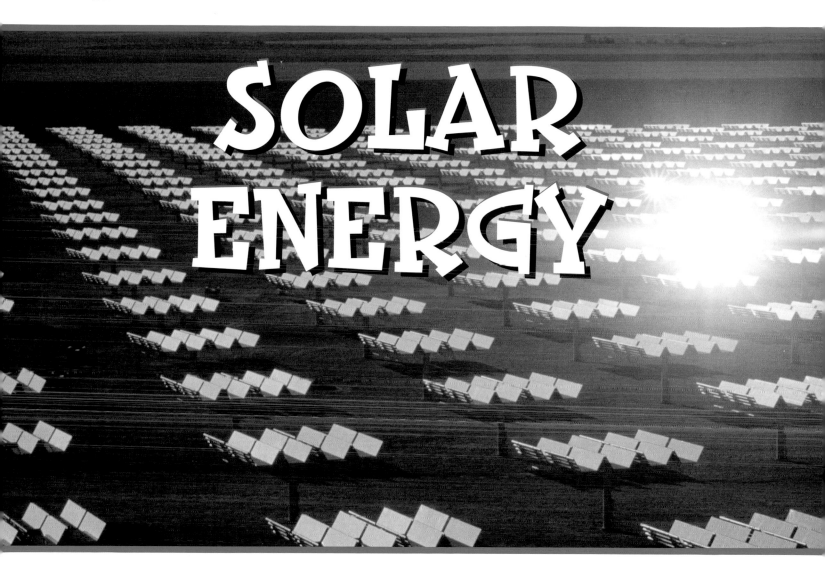

The Rosen Publishing Group's
PowerKids Press™
New York

Ian F. Mahaney

Published in 2007 by the Rosen Publishing Group, Inc.
29 East 21st Street, New York, NY 10010

First Edition

Editor: Joanne Randolph
Book Design: Julio Gil

Illustration p. 11 by Michelle Innes

Photo Credits: Cover, title page © Roger Ressmeyer/Corbis; p. 5 © PhotoDisc; p. 7 © Sie Productions/zefa/Corbis; p. 8 © www.istockphoto.com/Mary Marin; p. 9 © www.istockphoto.com/Carla Lisinski; p. 10 © Frank Krahmer/zefa/Corbis; p. 12 © Martin Harvey/Corbis; p. 13 © Galen Rowell/Corbis; p. 14 © José Fuste Raga/zefa/Corbis; p. 15, 17 © George Steinmetz/Corbis; p. 19 NASA; pp. 20, 21 Cindy Reiman; p. 22 Scott Bauer for The Rosen Publishing Group.

Library of Congress Cataloging-in-Publication Data

Mahaney, Ian F.
  Solar energy / Ian F. Mahaney.
      p. cm. — (Energy in action)
  Includes index.
  ISBN (10) 1-4042-3479-9 (13) 978-1-4042-3479-6 (lib. bdg.) —
ISBN (10) 1-4042-2188-3 (13) 978-1-4042-2188-8 (pbk.)
  1. Solar energy—Juvenile literature. I. Title. II. Energy in action (PowerKids Press)
  TJ810.3.M32 2006
  621.47—dc22
                                            2005036331

Manufactured in the United States of America

CONTENTS

The Sun	4
Solar Energy	6
Life and Solar Energy	8
How Plants Use Solar Energy	10
How Animals Use Solar Energy	12
Renewable Energy	14
Clean Electricity	16
Uses of Solar Energy	18
Experiments with Solar Energy	20
Glossary	23
Index	24
Web Sites	24

The Sun

The Sun is a powerful **source** of **energy**. Heat and light from the Sun warms Earth and makes life possible.

The Sun is a star much like the thousands of stars that we can see in the night sky. Stars are burning balls of gases that produce heat and light. Deep in the center of the Sun and other stars, many fast-moving **atoms** smash into one another. This creates a huge amount of heat and light. The heat and light created by the Sun **radiates**, or spreads out, in all directions. Plants and animals, including people, are able to use this heat and light in many ways.

The Sun is the closest star to Earth. Light from the Sun takes about 8 minutes to reach Earth. Light from the next nearest star takes about four years to reach Earth. Here we see a star-filled sky. Did you realize that many stars are bigger than our sun? The other stars look so much smaller because they are so far away.

Solar Energy

The biggest difference between the stars we see in the night sky and the Sun is the Sun's closeness to Earth. The Sun is only 93 million miles (150 million km) from Earth. That may seem like a huge distance, but we are close enough to benefit from the heat and light that the Sun radiates. On Earth we receive only a tiny amount of the heat and light created by the Sun. This small amount is still plenty to warm and light Earth.

The heat and light we receive from the Sun is energy called **solar energy**. Solar energy is used for many purposes that are important to life on Earth.

The Sun's light makes it possible for people, plants, and animals to live. It provides warmth and energy and allows plants to create food and people to create electricity. That is not bad for a star that is 93 million miles (150 million km) away!

Life and Solar Energy

Can you imagine what it would be like living on Earth if there were no Sun? It would be awfully dark and cold. It would be too dark and cold a place for people, other animals, and plants to live. There are certain conditions on Earth that make life possible. One condition is that water exists. Another is that light and heat are created by the Sun. People, along with all other animals and plants, need heat and light from the Sun to live. For example, there are animals that need help from the Sun to keep their bodies warm. These are cold-blooded animals. Without the Sun's warmth, these animals would die.

Opposite: Algae, such as the green algae here, needs sunlight to live. Algae are living things that often grow in water that does not move much, such as ponds and lakes. *Above:* Here an alligator suns itself on a rock in the Florida Everglades. You may see alligators or snakes lying on rocks in the sunlight. These are cold-blooded animals that use the heat from the Sun to warm their bodies.

How Plants Use Solar Energy

Most plants use energy from the Sun to make their own food. They do so through a **process** called **photosynthesis**. "Photosynthesis" means "made from light." Plants take in water and carbon dioxide through their roots and leaves. Carbon dioxide is a gas that is plentiful on Earth. The water and carbon dioxide combine with sunlight in the leaves to create glucose and **oxygen**. Glucose is a kind of sugar. Plants use this sugar for food. Plants let out the oxygen that is created during photosynthesis into the air. Animals use this oxygen to breathe.

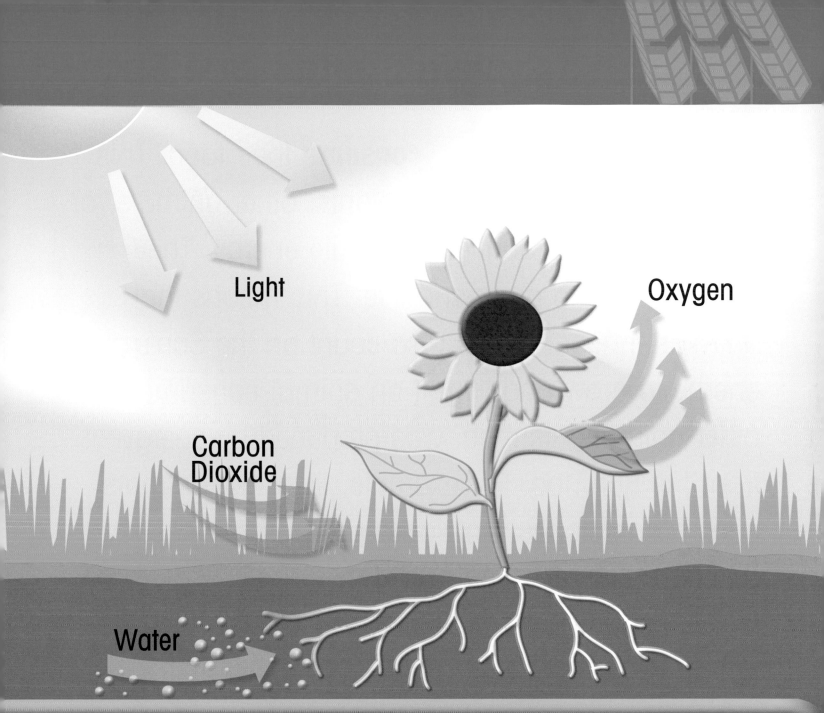

Light

Oxygen

Carbon
Dioxide

Water

Opposite: A plant is beginning to grow. It will grow well if it has good soil and can get enough sunlight and water. Most plants need at least some sunlight in order to live. *Above:* This picture shows how photosynthesis works. The plant takes in light and carbon dioxide through its leaves. It takes in water from the soil. Once the plant creates food, it gives off oxygen through its leaves.

How Animals Use Solar Energy

Animals, unlike plants, **consume** their food. This means that all animals eat plants, other animals, or both. Plants use sunlight to create energy. They need the Sun to grow and live. This means that animals that eat plants also count on the Sun's energy to grow and live. If an animal consumes other animals, its meal was produced by energy from the Sun, too. For example, lions eat animals, such as antelopes. Antelopes eat plants. Plants get their **nutrition** by using sunlight. Lions would not have food if it were not for antelopes, plants, and the Sun.

Opposite: Many animals, such as this zebra, eat grass or other plants. The plants need the Sun to grow and live, and these animals need the plants to stay healthy. Animals also need oxygen to stay alive. Plants put oxygen into the air as a result of photosynthesis. *Above:* This lioness hunts zebras on the grassy plains of Kenya. Without the Sun the lioness would not have zebras to eat.

Renewable Energy

People need and use solar energy daily, as plants and other animals do. We are a part of the **food chain**. This means we are as dependent on photosynthesis as any plant or other animal. We also use solar energy for other purposes. Making electricity is one way we use solar energy.

Most of the electricity that people make to use in our daily lives comes from **nonrenewable resources**. Oil is a nonrenewable resource.

There are many **renewable** sources of power on Earth, though. Solar energy is a renewable resource. Scientists **estimate** that the Sun will continue to shine for the next four to five billion years.

Opposite: Windmills use the power of the wind to create electricity or to grind grain. Wind is a renewable source of energy. *Above:* The Black Thunder Coal Mine in Wyoming is one of the largest suppliers of coal in the United States. Coal is a nonrenewable resource. Once all the coal is mined in this area, there will not be any more.

Solar energy is an endless source of energy for electricity. It is also a clean source of energy. When we make electricity from oil or coal, we burn them and the burning results in **by-products**. When you burn wood on a campfire, smoke rises in the air. Burning oil releases soot and gases that harm the air surrounding Earth. We can make electricity from solar energy without releasing by-products into the air. We create electricity from solar energy by allowing the Sun to heat water pipes that are covered by **solar panels**. Nothing is burned in the process. We call this a clean-air **alternative** in making electricity.

This is the Rancho Seco power plant. Since 1989, this plant has created most of its power using solar energy. The solar panels used to collect the sunlight are the flat rectangles you see here.

Uses of Solar Energy

We use solar energy for many purposes other than to produce electricity. Have you ever seen a solar-powered **calculator**? A solar-powered calculator gathers all the energy that it needs to operate by collecting light. We can also use solar panels to create energy in our homes. Many people in sunny places like Arizona, Florida, and California capture sunlight to make hot water for their homes or to produce their own electricity.

People are always experimenting with new **technology**, too. Can you imagine a solar-powered car? There are cars made that run on sunlight rather than on gasoline. Can you think of other ways that we use solar power in our lives?

The International Space Station uses solar panels to create some of its power. The solar panels are the long rectangles you can see in this picture. The International Space Station was put into orbit around Earth in 1998.

SUPPLIES NEEDED:

3 empty trash cans, large trash bags, or other sturdy containers, 3 pieces of construction paper, clear tape, a felt-tip marker, trash

Using solar energy is one way we can save Earth's resources. We can also recycle many of the things that we throw away as garbage. Recycling is when used things are broken down and made into new products. Most cities and towns have ways to recycle newspaper, glass, aluminum, and plastic. Try this project to start your own recycling center.

Step 1 Find out how garbage is recycled in your neighborhood. If it's like most places, they probably divide the recyclable trash into three groups: paper/cardboard, aluminum/glass, and plastic. Write each group name on a piece of paper.

20

Step 2 Tape one piece of paper to each container. These will be your recycling bins.

Step 3 Ask an adult if you can pick through the garbage can to pull out the recyclable things. Carefully remove all the paper products and put them in the paper bin. Remove the aluminum and glass and put them in their bin. Do the same thing with the plastic garbage.

Step 4 The next time you throw something away, decide in which bin it goes. Does it go in the garbage can or in one of the recycling bins? Encourage your family members to recycle, too.

Experiments with Solar Energy: Energy in Bonds

SUPPLIES NEEDED:
A 16-ounce plastic bottle, a cork that fits the bottle, baking soda, vinegar, bathroom tissue

Atoms often join with other atoms to make **molecules** and **compounds**. The places where these atoms connect are called bonds. There is a lot of energy stored in bonds. Solar energy is created when energy stored in bonds is released. Try this experiment to see this kind of energy in action.

Step 1 Pour about 3 ounces of vinegar into the bottle.

Step 2 Place about 1 teaspoon of baking soda on a tissue and fold it until it will fit inside the bottle. Put the tissue into the bottle and quickly cork the bottle. Make sure that the bottle is not pointing toward anyone, and stand back. What happens to the cork?

Glossary

alternative (ol-TER-nuh-tiv) A new or different way.

atoms (A-temz) The smallest parts of an element that can exist either alone or with other elements.

by-products (BY-prah-dukts) The results of burning fuel. Fuel is something that can be used to create energy.

calculator (KAL-kyuh-lay-ter) A machine or tool used to do math.

consume (kun-SOOM) To eat.

energy (EH-nur-jee) The power to do work or to act.

estimate (ES-tih-mayt) To make a guess based on knowledge or facts.

food chain (FOOD CHAYN) Living things connected because they are each other's food.

nonrenewable (non-ree-NOO-uh-buhl) Not able to be replaced once it is used.

nutrition (noo-TRIH-shun) The food that living things need to live and to grow.

oxygen (OK-sih-jen) A gas that has no color, taste, or odor and is necessary for people and animals to breathe.

photosynthesis (foh-toh-SIN-thuh-sus) The way in which green plants make their own food from sunlight, water, and carbon dioxide.

process (PRAH-ses) A set of actions done in a certain order.

radiates (RAY-dee-ayts) Spreads out from a center.

renewable (ree-NOO-uh-bul) Able to be replaced.

resources (REE-sor-sez) Things that occur in nature and that can be used or sold, such as gold, coal, or wool.

solar energy (SOH-ler EH-nur-jee) Heat and light created by the Sun.

solar panels (SOH-ler PA-nulz) Collectors used to capture and store solar energy.

source (SORS) The place from which something starts.

technology (tek-NAH-luh-jee) The way that people do something using tools and the tools that they use.

Index

A

atoms, 4

B

by-products, 16

C

clean-air alternative, 16

E

Earth, 4, 6, 8, 10, 14, 16

F

food chain, 14

H

heat, 4, 6, 8

L

light, 4, 6, 8, 10, 18

N

nutrition, 12

P

photosynthesis, 10, 14

S

scientists, 14
solar panels, 16

Web Sites

Due to the changing nature of Internet links, PowerKids Press has developed an online list of Web sites related to the subject of this book. This site is updated regularly. Please use this link to access the list: www.powerkidslinks.com/eic/solar/